YOUR KNOWLEDGE HAS VALUE

- We will publish your bachelor's and
  master's thesis, essays and papers

- Your own eBook and book -
  sold worldwide in all relevant shops

- Earn money with each sale

Upload your text at www.GRIN.com
and publish for free

Practical solution of selected geometrical problems by the iterative approximation method

Wladyslaw Chrusciel

Bibliographic information published by the German National Library:

The German National Library lists this publication in the National Bibliography; detailed bibliographic data are available on the Internet at http://dnb.dnb.de.

ISBN: 9783346838971
This book is also available as an ebook.

© GRIN Publishing GmbH
Nymphenburger Straße 86
80636 München

All rights reserved

Print and binding: Books on Demand GmbH, Norderstedt, Germany
Printed on acid-free paper from responsible sources.

The present work has been carefully prepared. Nevertheless, authors and publishers do not incur liability for the correctness of information, notes, links and advice as well as any printing errors.

GRIN web shop: https://www.grin.com/document/1339569

WLADYSLAW CHRUSCIEL is a doctor of technical sciences and computer scientist. He is also the author of books: *Information and Construction* and *The Evolution strategy of artifacts*.

For Anna & Robert
and for my entire Family, and its generations,
for whom science is also a way of thinking about the world.
My special thanks go to Olivia Johnston for reading my argumentation
'cover to cover', whose comments improved the text and helped me to maintain
the clarity and considered the tone of the message.
I also thank Natia Giorgidze the genius of creative discussions.

CONTENTS

PREFACE

> Logic changes fundamentally when we assume that besides truth and falseness, there is a third or more of such values.
>
> *Jan Lukasiewicz*

The book presents practical solutions for geometric problems using iterative approximation. But what is a geometric problem, and in particular its solution by iterative approximation?

Let's start with the etymology. Colloquially, a problem is a discrepancy between the actual state and a set or specified goal (target state) that cannot be solved routinely.

By using approximation (to Latin *proximus* - the next) in geometry, one can approach the correct solution. Complicated objects can often be approximated using polygons and circles.

Iteration is the repetition of a process by which, step by step, we approach a geometric solution until a satisfactory result is achieved. Each step of the process is a single iteration, and the outcome of each iteration is then the starting point of the next iteration.

In practice, effective problem solving involves both the appropriate determination of the scope and nature of the problem, its diagnosis in determining the goal, and the search for an optimal solution. It is a process of dynamic information processing. Dynamism, progress, and the search for solutions are associated with a constant increase in the amount of information, knowledge, and skills.

This means that information should not only be obtained but also transformed and used to create new knowledge that enables correct conclusions. Mathematics, including geometry, is a field of science in which, compared to the other sciences, experiments and inference play the most important role. All our knowledge comes from experimentation and inference, understood as the direct contact with reality. The more we want to know about reality, the more precise our thinking must be about the relation between geometric elements. The Greeks had one geometry, Euclid´s geometry, we have many.

I will not discuss or define the basic principles of Euclidean geometry, nor its origins, which are lost in the darkness of history, nor will I discuss its development. At the same time, I assume that the reader is familiar with elementary geometry. I will not discuss the use of drawing instruments or the making of geometric constructions, as I assume that these are known.

I merely want the reader to find a practical way to solve geometrical problems. At the same time, geometry should not only be considered a technician's, engineer's or scientist's field of work but as a toy that opens the way to an intellectually active mind, at the highest level.

The primary task of this work is to outline approximation methods for solving geometric constructions and to introduce them to the field of geometry. Knowledge of geometry greatly facilitates understanding. It also offers the possibility of graphically solving many vital problems that frequently occur in, for example, engineering and whose solution process would be inconvenient using computational methods.

This book is aimed at all those who are interested in the practical solution of geometric problems. The work consists of two parts, which are divided into chapters.

In the first part, which comprises chapters 1-3, I present a way of solving three problems of Euclidean geometry using only a compass and a ruler without a scale, which many mathematicians consider unsolvable - these are: (1) the trisection of any angle and any segment, (2) the quadrature of a circle, (3) the requirements for solving the doubling of a cube. However, these problems are not as simple as they might first appear. I am convinced that part of the difficulty is due to the fact that we are still looking at modernity from an outdated point of view. I would like to begin my observations by proposing to put an end to tiring traditional categorisation.

The second part, chapter 4, presents a way of thinking creatively using *chance*, which can provide us with useful knowledge when discovering something *new*.

The book concludes with a compilation of the literature used in this study.

South England, Bere Alston, early spring 2020

Wladyslaw Chrusciel

7

PART I

THE THREE OLDEST GEOMETRICAL PROBLEMS

1 – DIVISION OF ANY ANGLE INTO *N* EQUAL PARTS ONLY WITH COMPASS AND RULER

1.1 Introduction

> *This geometrical discovery was so surprising that the author initially doubted himself – but it turns out that the impossible is easier than we think.*
>
> *(Berlin, 29-April-2020, 10am, during breakfast)*

Throughout mathematical history, we have seen that ancient tradition could not solve the geometric construction problems related to the determinations of Euclidean geometry. The oldest three geometric problems are: squaring the circle, doubling the cube, and dividing the angle into three.

You can feel a certain resignation as there is no prospect of solving these geometrical problems.

Felix Klein wrote the following in his book [KLE24] on the question of unsolvable mathematical problems:

If something does not work in the usual way, one must not give up immediately and stick to the determination of the impossibility, but just find the right end at which the thing can be touched and further promoted. The mathematical thought as such never has an end, (...) [KLE24, p.154].

British mathematician Marcus du Sautoy and author of the book [DUS17] *What we cannot know* also mentions that:

Doubling the cube, squaring the circle, and a third classical problem of trisecting the angle all turned out to be impossible [DUS17, p. 374].

How can we understand an unsolvable geometric problem, if we *only* look at it through the lens of algebra?

A review of previous methods for dividing an angle into N equal parts shows that their authors [BRE51, BIB52, MEC60, DUS17, MIL86, PER17] formulate the geometric tasks abstractly and attempt to solve them purely using mathematical (algebraic) methods, not with geometric ones, which in turn leads to negative results. We know the problem's target state, but we do not know with which method this can be achieved.

Undoubtedly, the classical methods (bisection, trisection of an angle) prove to be a great help in certain cases, but they have the disadvantage of not being universally applicable in all cases. Therefore, an attempt was made to develop a uniform method for the geometric division of an arbitrary angle into N (2, 3, 4, 5..., n) equal parts. It will be shown that the approximation method is a helpful tool for dividing that can be used for all arbitrary angles.

1.2 Geometric constructions with compass and ruler without scale

> We cannot solve our problems with the same thinking we used when we created them.
>
> *Albert Einstein*

Using compasses and rulers in order to construct entails certain defined requirements that have their origins in ancient Greece and Euclidean geometry.

The question of why the Greeks came up with the idea to limit themselves to compasses and rulers, cannot be answered clearly. Only through interpretation of the postulates (1, 2, and 3) in Euclid's *Elements* [EUK10] can one presume their use of geometric tools (compasses and ruler). Literally speaking, compasses and rulers are not mentioned in Euclid's *Elements*. They are only metaphors for the first three Euclidean postulates. The tools do not really exist, they are merely formed through thought [FRO19].

When solving this problem, not only is a strict formalization of all geometric constructions and their comprehensive proofs required but all technical aids must be known to divide any angle into N equal parts (see Figure 1.1).

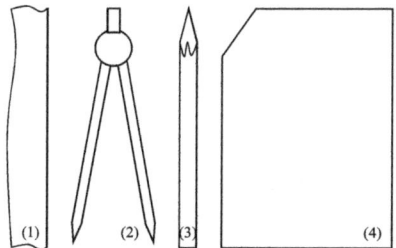

Figure 1.1 Aids: (1) – ruler without a scale (it is not possible to take measurements; since only one side of the ruler was used for drawing, it is not possible to draw straight parallels); (2) – compass; (3) – pencil 2H (4H); (4) – sheet of paper not lined or squared; The author assumes that the rules of geometric constructions and the use of tools are known to the readers

1.3 Trisection of the angle

In geometry, the *angle trisection* is understood as the problem of whether one can precisely subdivide any arbitrary angle into three equal parts with the help of only a compass and a ruler without a scale (the Euclidean tools). The angle trisection is one of the three classic problems of ancient mathematics and is currently only possible for certain angels, e.g. the 90° and 45° angle can be divided into three; it is also possible to divide the 135° and 180° angle into three parts (see Figure 1.2). And what about the trisection of the 60° angle or any other angle? The current answer is: *No! It really does not work* [MES60].

Nevertheless, there are many attempts to solve this problem (see online databases). For me, impossible is a relative term. I admire what was invented in the past. But apart from the past inventions, I wonder whether or not it was possible to think differently about constructions; using imagination to form structures based on empty metaphors and word games?

Figure 1.3 shows an interesting, but very complicated attempt to divide a 60° angle into three.

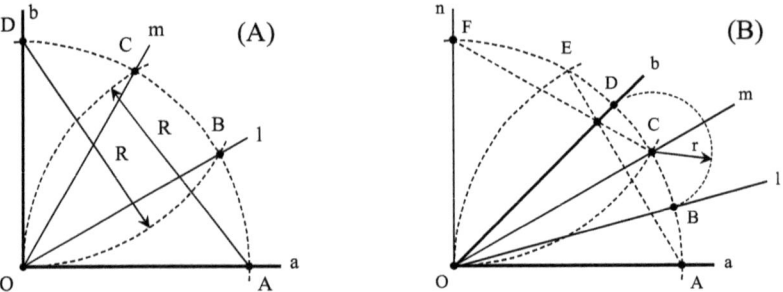

Figure 1.2 Trisection of the angle – classical method: (A) – ∠ *AOB* = 90° (∠ *AOC* = ∠ *COD* = ∠ *DOB* = 3(30°) = 90°); (B) – ∠ *AOB* = 45° (∠ *AOC* = ∠ *COD* = ∠ *DOB* = 3(15°) = 45°); Black points are only informative sings for the solution of the task and are irrelevant for the geometric construction

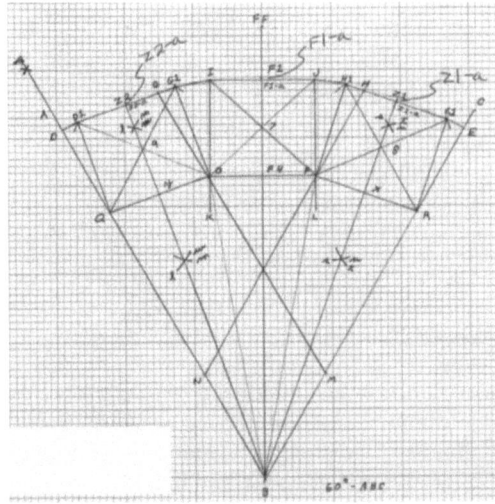

Figure 1.3 Trisection of the 60° angle [RED18] – but it is a solution on graph paper and therefore it is possible to make measurements (ruler with scale)

Let's analyse the current state of our knowledge. The problem of dividing an angle in half can be solved because we always use a known interpretation of geometric elements for this type of task. It is, however, essentially only aimed at division into an even number of equal parts. This process does not work to divide an arbitrary angle into exactly three equal parts using only a compass and a ruler without a scale (see Figure 1.4 (A)). The classic method is very static (for Euclid too, geometry is rigid and not flexible enough), meaning that one cannot just adopt the solution process of any other geometric problem, e.g. the division of an angle into an uneven number of equal parts, and expect it to work here too. That means we must change our point of view in order to solve the problem of dividing the angle.

Change in mindset – this is an old but still valid paradigm of innovation. This new geometric interpretation not only stimulates the imagination, but also uses analogies, associations and mechanisms of thought evolution. At the same time, new insights are created, which make it possible to achieve the desired goal. The problem should be looked at from within and from without [CHR20].

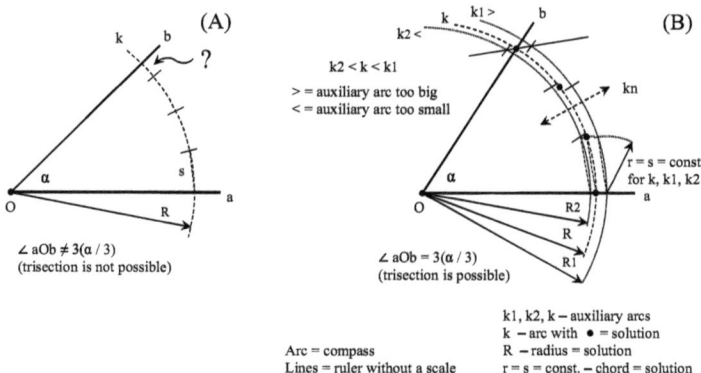

Figure 1.4 Trisection of any angle: (A) – Classical (static) method – trisection is not possible; (B) – Dynamic approach (we are looking for a family of arcs) – trisection is possible (the result is achieved in n working steps); Black points are only informative sings for the solution of the task and are irrelevant for the geometric construction

First and foremost, a new method of dividing an angle into an uneven number of equal parts must be designed in order to solve the problem at hand. It is crucial that the geometric construction process is not static as it was before (see Figure 1.4 (A)) but dynamic (see Figure 1.4 (B)).

To solve this task, certain steps need to be repeated several times until the necessary improvement is achieved. This is an iterative process by which the solution is approached step by step until the result appears satisfactory. The interconnections are complex and therefore the desired solution cannot be obtained in just one step. These iteration processes happen in all concretisation stages in the search for a solution to the problem [CHR20].

1.4 New determination of the method of dividing the angle

For every angle there is just one circular arc k that can be divided into exactly N equal parts. In order to solve this problem, a shared or common geometric element (feature) must be found, with which one can further experiment. The relations between these new elements can in turn be analysed. *The common and constant geometric element for the entire set of circular arcs is the chosen chord (s).* In other words, one must first find the entire set of circular arcs and then select the appropriate arc in connection to the chord (s).

The search for and representation of circular arcs combined with the appropriate chord is the most important aid in geometric construction for solving the task of dividing the angle with only a compass and a ruler. One can say that from a methodological point of view, the list of properties is supplemented by new geometric features. Based on executed examinations, it is possible to formulate geometric features that enable any angle to be divided into N equal parts. In the event of noncompliance, the construction should be improved by formulating new construction elements and choosing new primary terms.

Attention should always be paid to the correctness of the solution, to make sure not to create flawed constructions and mythological thoughts such as Platonic ideas. Instead, one should try to understand the essentials of the real world in which we live and act in. The aesthetic component in geometry has been a topic of discussion since Plato; it was discussed and analysed regarding blurred concepts such as order, proportion (divine proportion), balance, harmony, unity and clarity.

Man should examine geometric shapes through mental contemplation. How can one apply said standards into geometrical constructions? Beauty and aesthetic are always personal; they change according to culture and generation. Any emotional return to Greek thought will not only be of no benefit but also harmful.

In reality we don't know much about the early thinking of the Greeks, and what we do know is only fragmentary and of a controversial interpretation. This way of thinking was and is idealistic and has nothing to do with a practical way of thinking and acting. The lack of common features and relations between individual geometric elements makes it impossible to solve the problem. However, if we look at the problem from a different perspective, it may be solvable.

The geometrical constructions that we will solve are practical tasks. They differ in many ways from purely mathematical problems, but the main motives and methods of solving them are essentially the same. Practical construction tasks usually include mathematical tasks [POL10].

The solution of dividing any angle into N equal parts will be presented through two main tasks:

EXERCISE A – division of any angle into an *uneven number* of parts; Examples (A1, A2, A3, A4, A5, A6)

EXERCISE B – division of any angle into an *even number* of parts; Example (B1)

EXERCISE A – division of any angle into an *uneven number* of parts.
Formulation of the practical construction task:
    Divide any α angle into any odd number of equal parts (α / (3, 5, 7, 9)) only with the help of a compass and a ruler without a scale.

Our division problem will be solved if the angle is successfully divided into an odd number of equal units, and the beginning of the first and the endpoint of the last unit both lie on the arms of the given α angle (see Figure 1.4 (B)).
Labels:
Triangle: $\triangle ABC$
Straight lines: a, b, g, l
Circular arc; k1, k2, k; k1(O, R1) – the circular arc k1 around point O with radius R1;
$k = f(R)$ – the circular arc k is a function of radius R.
Lengths, distances: AB, E1E2
Points, intersections: A, B, C, C1, ..., E2′; solution intersections are often indicated by black points (are only informative signs for the solution of the problem and are irrelevant for the geometric construction); Intersections on auxiliary lines are indicated by small lines
Angle: α-angle, 45°- angle, angle aOb, $\angle$ aOb, $\angle$ AOC,
Numbers: $N = (2, 3, 4, 5, 6 ..., n)$;
Even numbers (2, 4, 6 ...);
Odd numbers (3, 5, 7, 9 ...);

EXERCISE A, Example (A1) is solved in steps.

EXAMPLE (A1) – trisection of 90° angle (see Figures 1.5 to 1.7)
What is known?
Given: the $α = 90°$ angle aOb,
The odd number of parts: (α/3),
Aids: compass, ruler without a scale

Scope of construction work – general remarks:
What is unknown?
    First and foremost, the method of dividing to be used in this geometrical construction is primarily unknown. The *approximation* method is used to formulate the *new method*.
    The incremental strategy is understood as a way to improve existing solutions, meaning it entails an evolutionary development of the geometric structure. This is a traditional, frequently used procedure.
    Even though we do not need any special knowledge to understand this task, I want to clearly display it. Figure 1.5 shows the idea for the solution and Figure 1.6 shows the main cases (F-1 to F-3) of the approximation constructions.

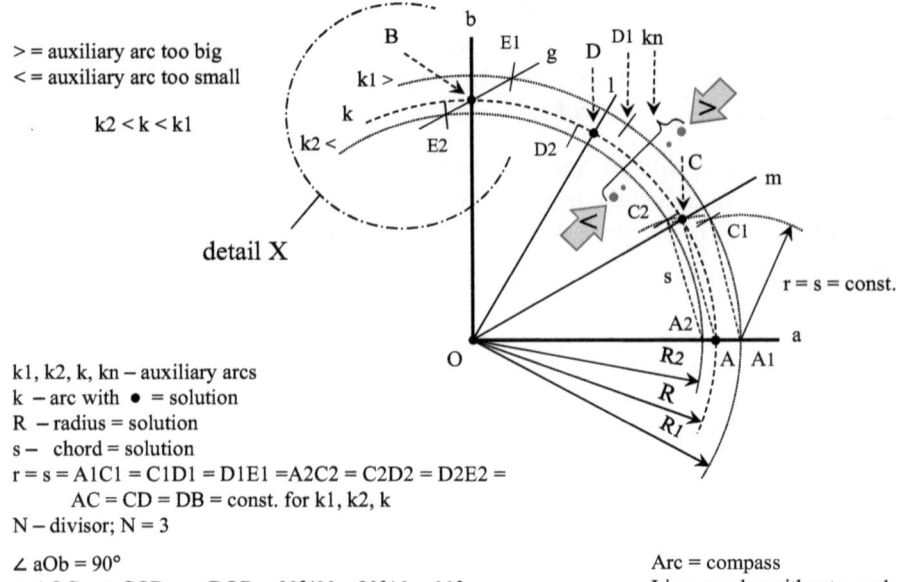

> = auxiliary arc too big
< = auxiliary arc too small

k2 < k < k1

detail X

k1, k2, k, kn – auxiliary arcs
k – arc with • = solution
R – radius = solution
s – chord = solution
r = s = A1C1 = C1D1 = D1E1 =A2C2 = C2D2 = D2E2 =
    AC = CD = DB = const. for k1, k2, k
N – divisor; N = 3

∠ aOb = 90°
∠ AOC = ∠ COD = ∠ DOB = 90°/ N = 90°/ 3 = 30°

r = s = const.

Arc = compass
Lines = ruler without a scale

Figure 1.5 Sketch of an idea for a solution: trisection of the 90°-angle (90°/ 3); Black points are only informative sings for the solution of the task and are irrelevant for the geometric construction

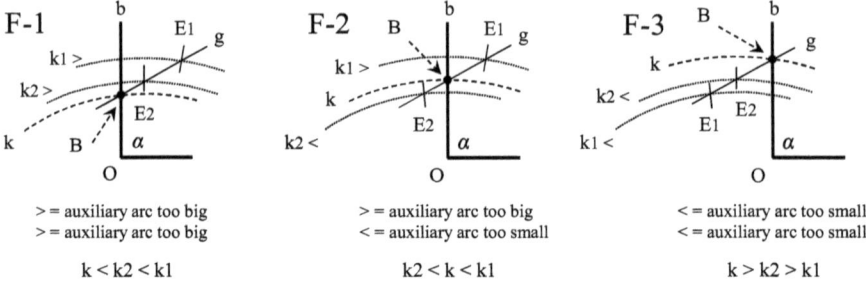

F-1
> = auxiliary arc too big
> = auxiliary arc too big
k < k2 < k1

F-2
> = auxiliary arc too big
< = auxiliary arc too small
k2 < k < k1

F-3
< = auxiliary arc too small
< = auxiliary arc too small
k > k2 > k1

Figure 1.6 Possible main cases (F-1 to F-3) of approximation constructions $kn = f(Rn)$ when dividing the angle: (see Figure 1.5 detail X); Black points are only informative sings for the solution of the task and are irrelevant for the geometric construction

14

Construction – Figure 1.7, trisection of the 90°-angle, steps (I) to (IV)

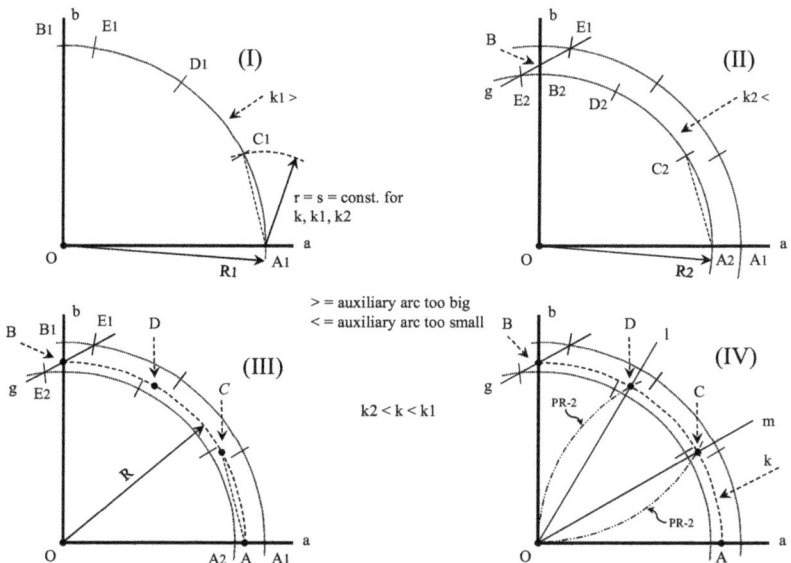

Figure 1.7 (I - IV) – Construction steps (I)-(IV); Step (IV) = Solution; Solution is marked with *A, B, C*; Next to solutions are marked with *A1, A2, B1, B2 ...*; **Trisection of the 90°**-angle *AOB* ($\angle AOC = \angle COD = \angle DOB = 30°$; 3(30°) = 90°); Radius $r = s = A1C1$ (chord) = *const.* for steps (I – IV); Solution: see note step (III)-(8); In step (IV) *PR-2* – proof (*PR-2* – the correctness of the result of the task solution can also be proven by the classic method of trisection of the 90°-angle); Black points are only informative sings for the solution of the task and are irrelevant for the geometric construction

Step (I) – Figure 1.7 (I)

(1) – draw the angle *aOb* = 90°,

(2) – with the radius *R1*, draw an auxiliary circular arc *k1* around *O*, which intersects the arms *a* and *b* in *A1* and *B1*,

(3) – divide the arc *k1* (*A1B1*) into 3 parts with any radius *r*; name the dividing points *C1, D1* and *E1*.

Remark:
The radius *r* (*r = s = const. = A1C1* (chord)) selected for a task cannot be changed and is used in all steps (I-IV).

Step (II) – Figure 1.7 (II)

(4) – draw an auxiliary circular arc *k2* with the radius *R2* around *O*, which intersects the arms *a* and *b* in *A2* and *B2*,

(5) – divide the arc *k2* into 3 parts with the chord *r = s = const.*, name the division points *C2, D2* and *E2*,

15

(6) – connect the end points *E1* and *E2* of the two circular arcs *k1* and *k2* with a segment *g*; the point of intersection with *b* is *B*.

Step (III) – Figure 1.7 (III)

(7) – draw an arc around *O* with the radius $R = OB$, which intersects the arm *a* at point *A*,

(8) – divide the arc *AB* into 3 parts with the radius $r = s =$ const., name the dividing points *C*, *D*, *B*.

Remark:

The greater the distance between the arcs *k1* and *k2*, the more arcs *kn* lead to the solution of the problem. However, as the number of arcs increase, the number of mistakes increase. Information overload often prevents effective interpretation. The structure for the solution is gradually developed further and further during the construction process.

The division is completed (solved) exactly when, and only when, the endpoints A and B of the arc k lie on both arms (a and b) of the given α-angle.

Step (IV) – Figure 1.7 (IV)

(9) – draw the lines *l* and *m* from *O* through *C* and *D*; the given angle $\alpha = 90°$ is divided into three equal parts.

What is the conclusion?

Theorem 1.1

The given α angle between the two arms *a* and *b* is divided into three equal angles by the straight lines *l* and *m*.

$\angle AOC = \angle COD = \angle DOB = (\alpha / 3) = 30°; 3(30°) = 90°$

Theorem 1.2

The congruence of distances is an equivalence relation.

$OA = OC = OD = OB = R$

$AC = CD = DB = s = r =$ const.

Theorem 1.3

Three triangles are congruent when the three sides are equal.

$\triangle AOC = \triangle COD = \triangle DOB$

Theorem 1.4

If three triangles are congruent, the corresponding angles are equal.

$\angle AOC = \angle COD = \angle DOB = (\alpha / 3) = 30°; 3(30°) = 90°,$

PROOF (PR-1):

If the endpoints A and B of the arc k lie on both arms a and b of the α-angle, then by construction, the triangles are congruent (see theorem 2.4) and the angle is thus precisely divided into three equal angles.

Which was to be proven.

PROOF (PR-2):

The solution's correctness can also be proven by the classic method of dividing the 90°-angle into three (Figure 1.7 (IV), proof *PR-2*).

EXAMPLE (A2) – trisection of 45°-angle (see Figure 1.8 (A))

What is known?

Given: the $\alpha = 45°$ angle aOb,

The odd number of parts: $(\alpha/3)$,

Aids: compass, ruler without a scale

Scope of construction work: as in Example (A1).

EXAMPLE (A3) – trisection of 60°-angle (see Figure 1.8 (B))

What is known?

Given: the $\alpha = 60°$ angle aOb,

The odd number of parts: $(\alpha/3)$,

Aids: compass, ruler without a scale

Scope of construction work: as in Example (A1).

PROOF:

If endpoints A and B of the arc k lie on both arms a and b of the α-angle, then by construction, the triangles are congruent (see theorem 1.4, Example (A1)) and the angle is thus precisely divided into 3 equal angles.

Which was to be proven.

EXAMPLE (A4) – trisection of $\alpha°$-angle (see Figure 1.9)

What is known?

Given: the $\alpha°$-angle aOb,

The odd number of parts: $(\alpha/5)$,

Aids: compass, ruler without a scale

Scope of construction work: as in Example (A1).

PROOF:

If endpoints A and B of the arc k lie on both arms a and b of the α angle, then by construction, the triangles are congruent (see theorem 1.4, Example (A1)) and the angle is thus precisely divided into 5 equal angles.

Which was to be proven.

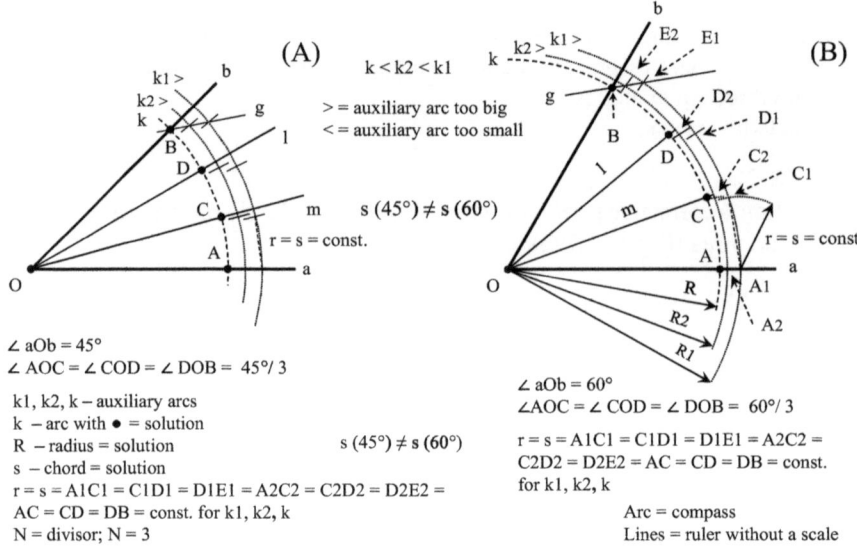

Figure 1.8 Trisection of the angle: (A) 45°/ 3; (B) 60°/ 3; Black points are only informative signs for the solution of the task and are irrelevant for the geometric construction

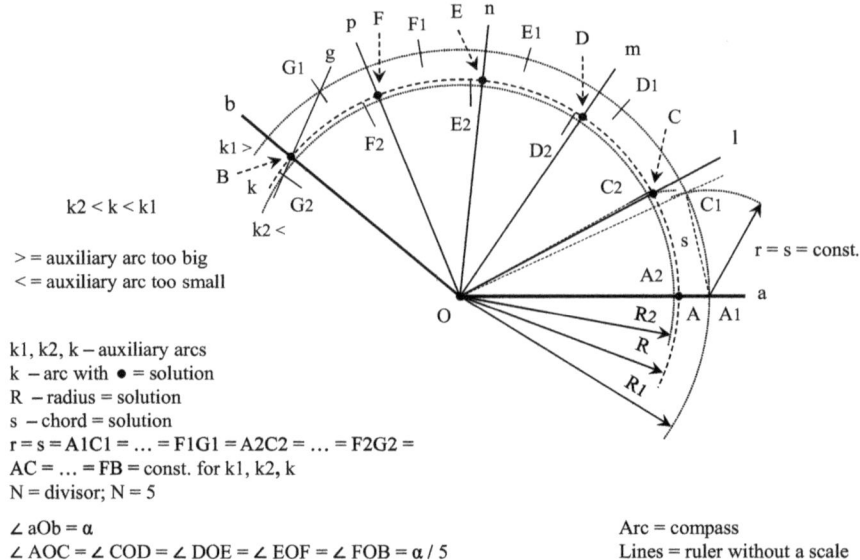

Figure 1.9 Division of the α-angle into 5 equal parts (α°/5); Black points are only informative signs for the solution of the task and are irrelevant for the geometric construction

18

EXAMPLE (A5) – division of the α-angle into 7 equal parts (see Figure 1.10)
What is known?
Given: the α angle *aOb*,
The odd number of parts: (α/7),
Aids: compass, ruler without a scale
Scope of construction work: as in Example (A1)

PROOF:
If endpoints *A* and *B* of the arc *k* lie on both arms *a* and *b* of the α-angle, then by construction, the triangles are congruent (see theorem 1.4, Example (A1)) and the angle is thus precisely divided into 7 equal angles.
Which was to be proven.

EXAMPLE (A6) – division of the α-angle into *3* and *9* equal parts (see Figure 1.11)
What is known?
Given: the α angle *aOb*,
The odd number of parts: (α/3 and α/9),
Aids: compass, ruler without a scale
Scope of construction work: as in Example (A1)

PROOF:
If the endpoints *A*, *A′* and *B*, *B′* of the arcs *k*, *k′* lie on both arms *a* and *b* of the α-angle, then by construction the triangles are congruent (see Theorem 1.4, Example (A1)) and the angle is thus precisely divided into 3 and 9 equal angles.
Which was to be proven.

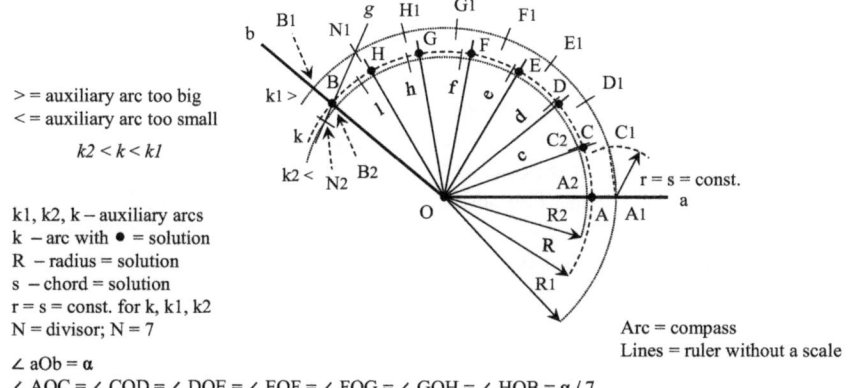

> = auxiliary arc too big
< = auxiliary arc too small

$k2 < k < k1$

k1, k2, k – auxiliary arcs
k – arc with ● = solution
R – radius = solution
s – chord = solution
r = s = const. for k, k1, k2
N = divisor; N = 7

∠ aOb = α
∠ AOC = ∠ COD = ∠ DOE = ∠ EOF = ∠ FOG = ∠ GOH = ∠ HOB = α / 7

r = s = const.

Arc = compass
Lines = ruler without a scale

Figure 1.10 Division of the α-angle into 7 equal parts (α / 7); Black points are only informative sings for the solution of the task and are irrelevant for the geometric construction

19

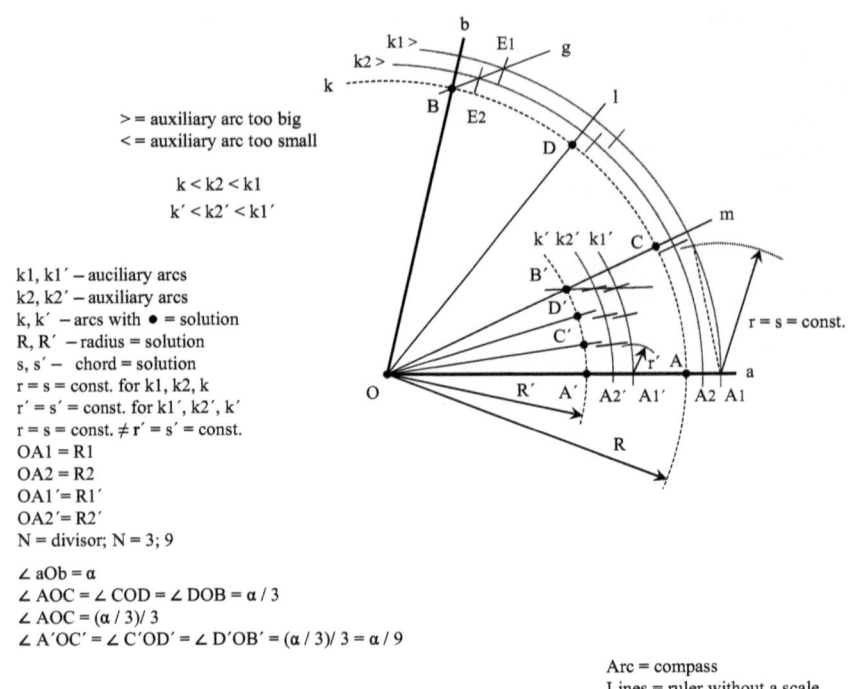

> = auxiliary arc too big
< = auxiliary arc too small

$k < k2 < k1$
$k' < k2' < k1'$

k1, k1' – auciliary arcs
k2, k2' – auxiliary arcs
k, k' – arcs with ● = solution
R, R' – radius = solution
s, s' – chord = solution
r = s = const. for k1, k2, k
r' = s' = const. for k1', k2', k'
r = s = const. ≠ r' = s' = const.
OA1 = R1
OA2 = R2
OA1' = R1'
OA2' = R2'
N = divisor; N = 3; 9

∠ aOb = α
∠ AOC = ∠ COD = ∠ DOB = α / 3
∠ AOC = (α / 3)/ 3
∠ A'OC' = ∠ C'OD' = ∠ D'OB' = (α / 3)/ 3 = α / 9

Arc = compass
Lines = ruler without a scale

r = s = const.

Figure 1.11 Division of the α-angle into 3 equal parts (α / 3) and 9 equal parts (α / 3) / 3 = α / 9; Black points are only informative signs for the solution of the task and are irrelevant for the geometric construction

General remarks on dividing any α-angle (see Figure 1.12). Amount of construction work: as in Example (A1).

It is not possible to measure individual geometric elements directly if one only uses the Euclidean tools. On the other hand, by introducing an algebraic mathematical apparatus into Euclidean geometry and by analysing the trisection of any angle in the Cartesian coordinate system, it is possible to accurately determine the relations between the various geometric elements. The additional lines, called auxiliary lines, complicate the geometric construction but form an essential part of the deduction process. They reorganise the construction into subconstructions, and it is precisely on this sublevel that the argumentation of the proof takes place (see Figure 1.12).

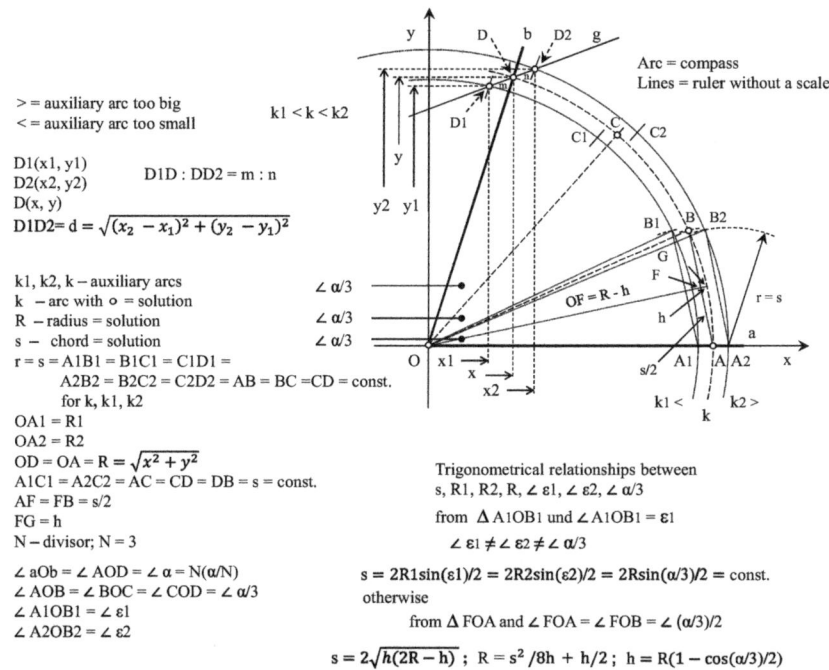

Left side text (labels):

> = auxiliary arc too big
< = auxiliary arc too small

$k_1 < k < k_2$

$D_1(x_1, y_1)$
$D_2(x_2, y_2)$
$D(x, y)$

$D_1D : DD_2 = m : n$

$D_1D_2 = d = \sqrt{(x_2 - x_1)^2 + (y_2 - y_1)^2}$

k_1, k_2, k – auxiliary arcs
k – arc with o = solution
R – radius = solution
s – chord = solution
$r = s = A_1B_1 = B_1C_1 = C_1D_1 =$
$\quad A_2B_2 = B_2C_2 = C_2D_2 = AB = BC = CD = $ const.
$\quad$ for k, k_1, k_2
$OA_1 = R_1$
$OA_2 = R_2$
$OD = OA = R = \sqrt{x^2 + y^2}$
$A_1C_1 = A_2C_2 = AC = CD = DB = s = $ const.
$AF = FB = s/2$
$FG = h$
N – divisor; $N = 3$

$\angle aOb = \angle AOD = \angle \alpha = N(\alpha/N)$
$\angle AOB = \angle BOC = \angle COD = \angle \alpha/3$
$\angle A_1OB_1 = \angle \varepsilon_1$
$\angle A_2OB_2 = \angle \varepsilon_2$

Right side text:

Arc = compass
Lines = ruler without a scale

$\angle \alpha/3$
$\angle \alpha/3$
$\angle \alpha/3$

$OF = R - h$

$r = s$

Trigonometrical relationships between
$s, R_1, R_2, R, \angle \varepsilon_1, \angle \varepsilon_2, \angle \alpha/3$

from $\triangle A_1OB_1$ und $\angle A_1OB_1 = \varepsilon_1$

$\angle \varepsilon_1 \neq \angle \varepsilon_2 \neq \angle \alpha/3$

$s = 2R_1\sin(\varepsilon_1)/2 = 2R_2\sin(\varepsilon_2)/2 = 2R\sin(\alpha/3)/2 = $ const.
otherwise
from $\triangle FOA$ and $\angle FOA = \angle FOB = \angle (\alpha/3)/2$

$s = 2\sqrt{h(2R - h)}$; $R = s^2/8h + h/2$; $h = R(1 - \cos(\alpha/3)/2)$

Figure 1.12. Trisection of any α-angle in the Cartesian coordinate system and the relations between the various geometric elements; Black points (zero circles) are only informative signs for the solution of the task and are irrelevant for the geometric construction

EXERCISE B – dividing any angle into an *even* number of parts.

Formulation of the practical construction task:
Divide any α angle into an even number of equal parts (α / 2) only with the help of a compass and a ruler without a scale.

EXAMPLE (B1) – Double division of the 90°-angle (see Figure 1.13)
What is known?

Given: the $\alpha = 90°$ angle aOb,
The *even number* of parts: (α/2),
Aids: compass, ruler without a scale
Scope of construction work: as in Example (A1)

PROOF (PR-1):
   If the endpoints A and C of the arc k lie on both arms a and b of the α angle, then by construction, the triangles are congruent (see theorem 1.4, Example (A1)) and the angle is thus precisely divided into 2 equal angles.
Which was to be proven.

PROOF (PR-2):

The solution's correctness can also be proven by the classic method of dividing the 90°-angle into two (see Figure 1.13 and proof *PR-2*).

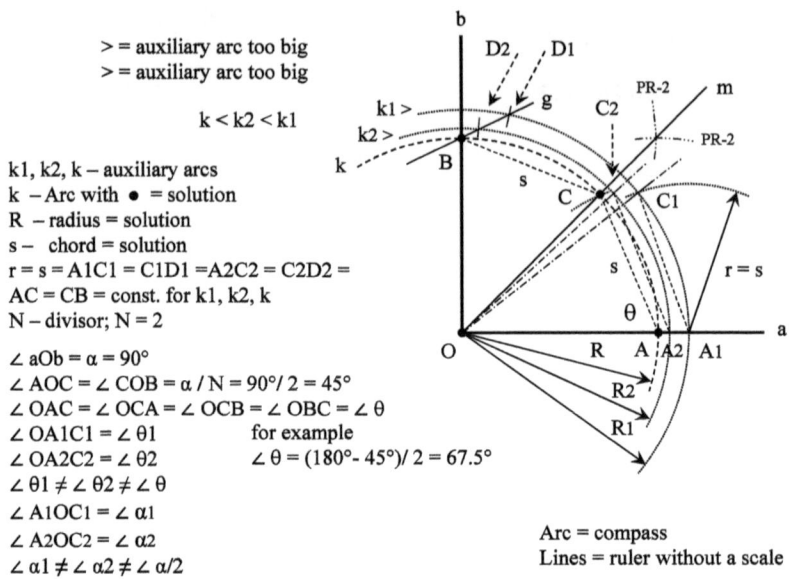

> = auxiliary arc too big
> = auxiliary arc too big

$k < k2 < k1$

k1, k2, k – auxiliary arcs
k – Arc with ● = solution
R – radius = solution
s – chord = solution
r = s = A1C1 = C1D1 = A2C2 = C2D2 =
AC = CB = const. for k1, k2, k
N – divisor; N = 2

$\angle aOb = \alpha = 90°$
$\angle AOC = \angle COB = \alpha / N = 90°/2 = 45°$
$\angle OAC = \angle OCA = \angle OCB = \angle OBC = \angle \theta$
$\angle OA1C1 = \angle \theta1$
$\angle OA2C2 = \angle \theta2$ for example
$\angle \theta1 \neq \angle \theta2 \neq \angle \theta$ $\angle \theta = (180° - 45°)/2 = 67.5°$
$\angle A1OC1 = \angle \alpha1$
$\angle A2OC2 = \angle \alpha2$
$\angle \alpha1 \neq \angle \alpha2 \neq \angle \alpha/2$

Arc = compass
Lines = ruler without a scale

Figure 1.13 Division of the des 90° angle into 2 equal parts (*90°/2*); *PR-2* – Proof (*PR-2* – the correctness of the results of the task solution can also be proven by the classic method of dividing the 90° angle into two); It should be noted that only R, k, s, θ and the relations between these elements are a solution to any divide the angle into N equal parts; Black points are only informative signs for the solution of the task and are irrelevant for the geometric construction

The search range for the arcs should not be too vast. The geometric interpretation of the approximation offers a multitude of possibilities for the analysis of the construction task and, at the same time, allows for insight into the type of task, in particular into the nature of the conflict between contradicting requirements.

Our aim is to identify one arc as the one which does not encompass any geometrical or logical flaws and meets all the problem's requirements.

As you know, on-paper-designed geometric constructions are burdened with inaccuracies (errors). They are mainly the result of inaccuracies in the drawing tools. Despite this, the results of carefully conducted constructions are actually quite sufficient in practice and are generally not inferior to calculation methods. Unfortunately, the study of basic geometric construction has become out of fashion. Today, the use of CAD-2D technologies not only speeds up the drawing process of geometric structures but also allows one to experiment and discover new solutions in geometry, which is a vital element in shaping the *spatial imagination*. Computer-based, innovative, and experimental works in the field of geometry are widely used in practice, such as cartography, photogrammetry, and technical drawing. Investigating the properties of geometrical compositions also gives the possibility of solving problems that can only be solved with the use of geometry methods. There is no one-size-fits-all method or strategy for solving problems [POL10]. One must avoid using just one method or solution strategy in geometry; adding new aspects to our methods is critical.

The new, presented iterative approximation method enables the division of any angle into N = (2, 3, 4, 5, 6, 7..., n) equal parts, and the division values agree with other classical methods satisfactorily.

A completely different geometric solution is to divide a distance into three equal parts. The task of dividing a distance using only a compass and ruler without a scale is also about finding a suitable division method, more specifically, the method of drawing parallel lines.

A practical solution to the geometric construction is described in exercise C.

EXERCISE C – Dividing a line into three equal parts.
Formulation of the practical construction task:
A given distance is to be divided into three equal parts using only a compass and a ruler without a scale.

Example (C1) – Trisection of a line (see Figure 1.14)
What is known?
The distance AB,
The number of parts: $(N = 3)$,
Aids: compass, ruler without scale,
The method of dividing a section AB using only a compass and ruler is unknown.

Steps (I-III) - Figure 1.14 (I-III).
(1) - the distance AB is given (see Figure 1.14 (I)),
(2) - from point A draw a line a at any angle to AB (see Figure 1.14 (II)),
(3) - on line a, which starts from point A, measure $n = 3$ equal parts using a compass,
(4) - the end point E is connected to the point B by a straight line,
(5) - from points C and D lines are drawn parallel to EB which intersect the line AB at points C2 and D2 (see Figure 1.14 (III)).
(6) - draw an arc of radius $R1 = BE$ from point A and from point B an arc of radius $R2 = AE$ which intersect at point A1 (see Figure 1.14 (III)).
(7) - draw the segments $A1C1 = C1D1 = D1B$ on line $BA1$ with the compass, which are equal to the segments $AC = CD = DE$,
(8) - connect the points C and $C1$, D and $D1$; they intersect the line AB at the points C2, D2, which determine equal segments $AC2 = C2D2 = D2B$ - which was to be proven.

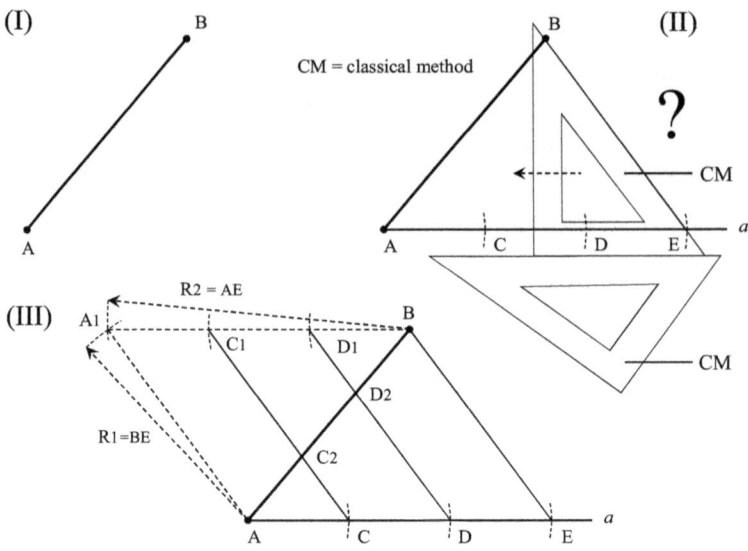

Figure 1.14 (I) A given distance *AB*; (II) – classical method (drawing parallel with the use of two triangles); (III) – trisection of a distance *AB* only with compass and ruler without scale; Black points are only informative signs for the solution of the task and are irrelevant for the geometrical construction

Remark:

Drawing parallel lines only with the help of compasses and rulers without a scale is a difficult problem to solve if one omits the classical method, i.e. the use of two triangles (see Figure 1.14 (II)). But this task can be solved e.g. with the method of drawing parallel lines as shown in Figure 1.14 (III).

Basic geometric constructions play an essential role, especially in descriptive geometry and technical drawing. Regardless of its practical application, geometry strengthens the imaginative power of spatial relations, and this is a valuable skill in the overall education of every technician and engineer. The geometric constructions shown can be used without a computer. Currently, however, geometry is usually supported by computer systems. The computed associated technique of constructing using specific equipment (plotters, computers) and programmes is referred to as CAD (Computer Aided Design). Geometry has added computers and plotters to the compass and ruler as versatile construction tools. With the help of dynamic geometry software, it is now quite easy to demonstrate exact mathematical proof. The current state of computer-associated geometry and construction contains a wealth of interesting and worthwhile application examples, which, however, are not the subject of this book.

2 – SQUARING THE CIRCLE

2.1 Introduction

The squaring of a circle is the geometric problem of constructing a square with area equal to the area of a given circle (see Figure 2.1 (A)). The problem is unsolvable since the number ($\pi = 3.1415926535\ldots$) is cannot be complied with. Using only the circle and the ruler, we can only determine a segment whose length is approximately equal to that of the circle. This construction is very important and often used in the practice of constructing artefacts. One of the most accurate and at the same time simplest methods is the one given by Polish mathematician Adam Kochanski in 1685 [GSE87].

The author will not discuss how to use drawing instruments and how to perform elementary geometrical constructions, e.g. drawing parallel and perpendicular lines. It is assumed that these already are known.

2.2 Kochanski's Approximation

We will first determine geometrically the approximate length of the given circle (see Figure 2.1 (B))

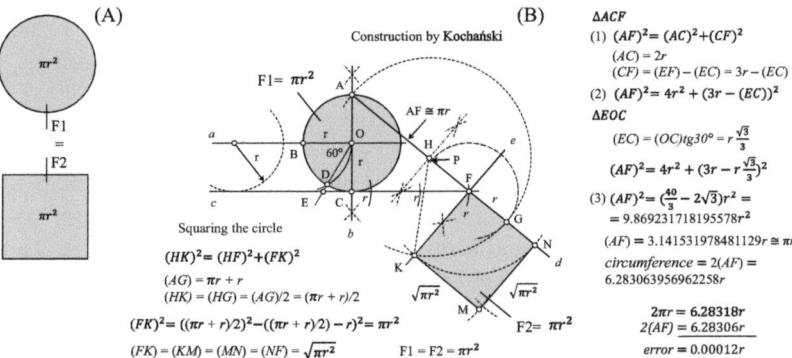

Figure 2.1 (A) – squaring the circle $F1 = F2$; (B) – geometric solution of the task, its algebraic justification and the value of the error made; Points (zero circles) are only informative signs for the solution of the task and are irrelevant for the geometric construction

We draw two equal circles of radius r on the straight line a. We draw a perpendicular line b through O, which marks points A and C. Then we draw a line c tangent to the two circles and passing through point C. From point B, we draw an arc with radius r, giving point D. The line through O and D intersects the line c at E, making an angle of $30°$ with the diameter AC. At the tangent c, we set off EF equal to three radii r of the circle. We draw a line d through points A and F. The segment AF is approximately equal to the

25

length of half the circle. After calculating the approximation error, it is equal to 0.00012 of the radius. For a circle with a diameter of e.g. 2 metres, the error is 12 hundredths of a millimetre, i.e. it is visually imperceivable.

2.3 Squaring problem

Based on the length of the half-circle (circumference/2), we determine the sides and the area of the square.

From point F we draw an arc with radius r which intersects line d at points G and P. Through point F we draw a perpendicular line e using points G and P. We divide the segment AG in half to obtain point H. From point H we draw an arc with radius $HG = AH$ which intersects line e at point K. From point F we draw an arc with radius FK which intersects line d at point N. From points K and N we draw arcs of radius FK intersecting at point M. The area $F1$ of the square $FKMN$ and the area $F2$ of the circle have the same value ($F1 = F2 = \pi r^2$). Figure 2.1 (A) presents the quadrature problem of a circle. Figure 2.1 (B) illustrates the geometric solution to the problem, its algebraic justification and the value of the error made.

3 – DOUBLING THE CUBE

3.1 Introduction

The cube doubling problem limits itself to the geometric construction of a cube whose volume $V2$ is twice the given $V1$ ($V2 = 2(V1)$). In algebraic terms, it is the geometric construction of the third-degree root of the number two. The classical solution to the problem using only a compass and a ruler without a scale is not possible. There are many solutions to this problem. One of them, due to its simplicity, is Isaac Newton's solution. In the following, I will only sketch the individual steps of a practical solution to this problem in general terms, with the important note *that only one of the geometric dimensions (a) is marked on the ruler* (Figure 3.1 (B)).

3.2 Practical geometric construction of the third degree root of two

We will first geometrically determine the length of one of the sides of the cube. We draw a straight line m through point A of the side of the cube. From the points A and B of the given side of the cube we draw arcs with radii $AB = a$, which intersect in the point F. Through the points F and A we draw a straight line n, which makes a 30° angle with the line m. The size of the edge $a = CD$ is marked on the ruler (R) with point C. While drawing segment BD, arrange the edge of the ruler (R) in such a way that it connects point *B* with point *C* of the ruler, which must lie on line *m*, and point *D*, which must lie on line *n*. The resulting *BC* is the sought edge *b* of the side of the doubled cube. Figure 3.1 (A) shows the cube doubling and Figure 3.1 (B) illustrated the geometric solution of the problem and its algebraic justification (1-9).

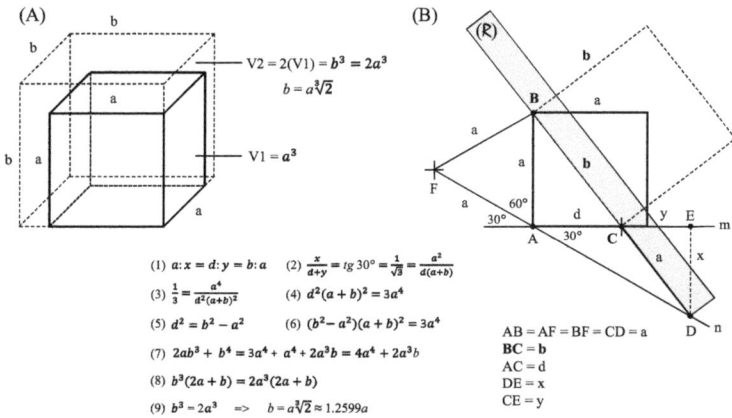

(1) $a:x = d:y = b:a$ (2) $\frac{x}{d+y} = tg\,30° = \frac{1}{\sqrt{3}} = \frac{a^2}{d(a+b)}$

(3) $\frac{1}{3} = \frac{a^4}{d^2(a+b)^2}$ (4) $d^2(a+b)^2 = 3a^4$

(5) $d^2 = b^2 - a^2$ (6) $(b^2-a^2)(a+b)^2 = 3a^4$

(7) $2ab^3 + b^4 = 3a^4 + a^4 + 2a^3b = 4a^4 + 2a^3b$

(8) $b^3(2a+b) = 2a^3(2a+b)$

(9) $b^3 = 2a^3$ => $b = a\sqrt[3]{2} \approx 1.2599a$

AB = AF = BF = CD = a
BC = b
AC = d
DE = x
CE = y

Figure 3.1 (A) Doubling the cube; (B) Geometric proof with algebraic calculations; Black dots are only informative signs for the solution of the task and are irrelevant for the geometric construction

3.3 Doubling the cube

The cube doubling is shown in Figure 3.2, which is based on the given geometric sizes of cubes *V*1 and *V*2.

The author does not go into the use of drawing instruments and the performance of elementary geometric constructions, e.g. drawing parallel and perpendicular lines, dividing a section and an angle into *N* equal parts. It is assumed that these are known.

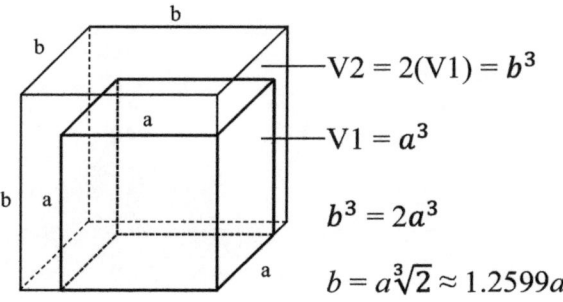

$V2 = 2(V1) = b^3$

$V1 = a^3$

$b^3 = 2a^3$

$b = a\sqrt[3]{2} \approx 1.2599a$

Figure 3.2 Doubling the cube $V2 = 2(V1)$

PART II

CHANCE IN CREATIVE THINKING IN THE GEOMETRY

4 – CONSTRUCTION AND CALCULATION OF NON-STANDARD GEOMETRICAL FORMS

4.1 Introduction

> Changing the landscape of human thought is not only the product of great minds, but also of humble artisans.
>
> *N.N.*

Problem solving is often described as a process of *chance*. It happens unexpectedly, it could not have been predicted. *Chance* in creative thinking can provide us with extremely useful knowledge in discovering something new.

So in this section I would like to present the practical side of *chance* in creative thinking.

4.2 The elliptic-circular rhombus (ECR)

The construction and calculation of bodies with non-standard shapes (see Figures 4.1 and 4.2) causes many difficulties in associating a spatial three-dimensional image to its record in the plane of rectangular projection. Tasks of this kind are difficult and require a lot of spatial three-dimensional imagination, especially if one uses the classical method of geometric drawing (compass and ruler). There are also difficulties in the production of these objects.

Figure 4.1 (A) Quadrund – Patent No. DE39917218/39917218.1, 1999, [NEI99]

CASE ANALYSIS: There is a patent on a certain geometrical solid called the *Quadrund* [NEI99], which is formed by interpenetrating two cylinders of the same diameter perpendicular to one another (Figure 4.1 (A)). In the context of this solution, I asked a simple question - what do we get when two straight rotating cylinders of the same diameter and length L penetrate and are tilted towards one another at a 45° angle ($\pi/4$), (Figure 4.1 (B)).

The new, accidental discovery was built on an already existing example. The process, where new ideas, new systems are superimposed on old ones (rather than constructed from scratch), is called *progressive technological overlap*.

The search result for similar solids in online databases turned out to be negative. The author named this new geometric object (Figures 4.1 (B) and 4.2) elliptic-circular rhombus (ECR).

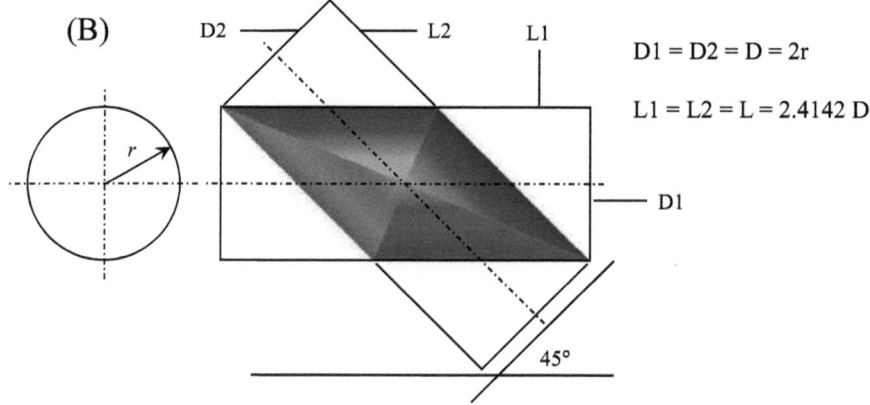

Figure 4.1 (B) The elliptic-circular rhombus (ECR); 3D models (A, B) were created by the author using CAD-3D Cocreate® © Wladyslaw Chrusciel

4.3 Volume and surface area – Cavalieri-Method

Francesco Cavalieri's principle was used to calculate the volume and area of an elliptic-circular rhombus. In its original formulation, it states that:

If two solids share the characteristic that their cross sections run though all planes, which are parallel to one predetermined plane, and have the same area, then these solids have equal volumes.

Figures 4.2 and 4.3 show the geometry and mathematical formulae required to calculate the volume of a solid as well as its area.

Properties of the new solid - the elliptic-circular rhombus:
- in the main view it is a rhombus, in the side view as well as at angle $\pi/4$ a circle and in the top view an ellipse,

- the sum of all interior angles of a rhombus is 2π (360°), the sum of the measures of its two neighbouring interior angles is π (180°), and the sum of the measures of its two acute interior angles is $\pi/2$ (90°),
- the diagonals intersect at a right angle, thus dividing the rhombus into four congruent right-angled triangles,
- the point of intersection of the diagonals divides each of them into halves, thus determining the centre of the inscribed circle, simultaneously being the rhombus' centre of symmetry,
- the diagonals coincide with the angle bisectors and the rhombus' axes of symmetry,
- in cross-section, along the diagonal f of the rhombus (see Figure 4.2 (A)), an ellipse is obtained whose major axis is equal to the diagonal f, and whose minor axis is equal to the diameter of the cylinders $D=2r$,
- in the section along the diagonal d of the rhombus (see Figure 4.2 (A)), an ellipse is obtained whose major axis is equal to the diagonal d, and the minor axis is equal to the diameter of the cylinders $D=2r$.

Geometrical relationships of the elliptic-circular rhombus elements

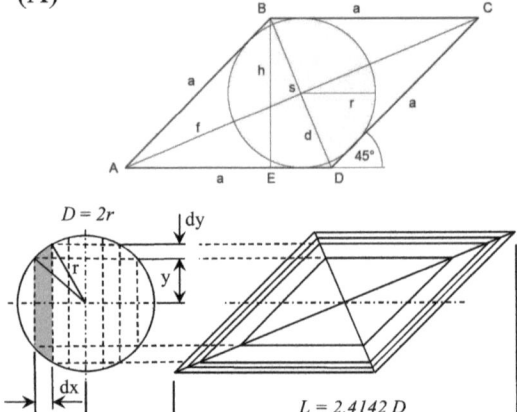

(A)

The height of the rhombus $ABCD$

$$h = 2y \qquad h = 2\sqrt{(r^2 - x^2)}$$

Side length

$$a = \sqrt{2}h$$

The scope of the rhombus ABCD

$$s = 4a$$

Section area $ABCD$

$$A = a \cdot h = \sqrt{2}h^2 = 4\sqrt{2}\,(r^2 - x^2)$$

Volume of the elliptic-circular rhombus

$$V = 4\sqrt{2} \cdot \int_{-r}^{r} (r^2 - x^2)\, dx = \frac{16}{3}\sqrt{2}r^3$$

Surface of the elliptic-circular rhombus

$$S = \frac{dV}{dr} = 16\sqrt{2}r^2$$

$L = 2.4142\,D$

Figure 4.2 (A) Calculation of the surface area and volume of an elliptic-circular rhombus (the principle of the Italian mathematician Francesco Bonaventura Cavalieri (1598-1647) was used to calculate the volume (ECR))

30

(B)

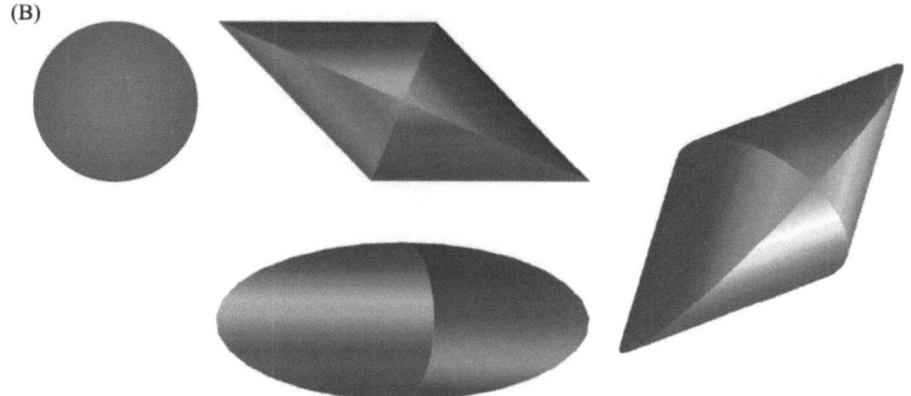

Figure 4.2 (B) The elliptic-circular rhombus (ECR) – CAD-3D Cocreate® © Władysław Chrusciel

4.4 Unfolding of the outer surface of the elliptical-circular rhombus

The result of unfolding the outer surface of the elliptic-circular rhombus (3D), the flat figure, is shown in Figure 4.3. The geometrical construction of the unfolding is very simplified.

Designing non-standard geometric shapes using the classical method with compass and ruler is very tedious and labour-intensive. It requires a lot of experience and concentration of the designer. The use of computer-aided CAD-3D not only facilitates the construction of this type of object, but also enables experimentation and the discovery of new forms in geometry, which is a very important element in shaping the designer's "spatial imagination". It is also about finding new mathematical facts. Computer-assisted innovation and testing in this field also offers the opportunity to observe the interplay between design and manufacturing technology. Practical applications: Aesthetic experience, especially where new forms of artistic expression are sought, e.g. in industrial design, arts, furniture and lighting.

In practice, it is often necessary to represent the development of the entire surface of a rotating (3D) figure. To represent the unwinding of the surface, it is sufficient to geometrically analyse only one half of the cylinder. A typical example is depicted in Figure 4.4.

The Socratic Method states that it is not actually the final conclusions that play a major role, as they are only temporary. It merely shows the current state of knowledge we are able to perceive and solve problems through. If further questions are posed, our conclusions would rapidly change, which would destroy the momentary order of things.

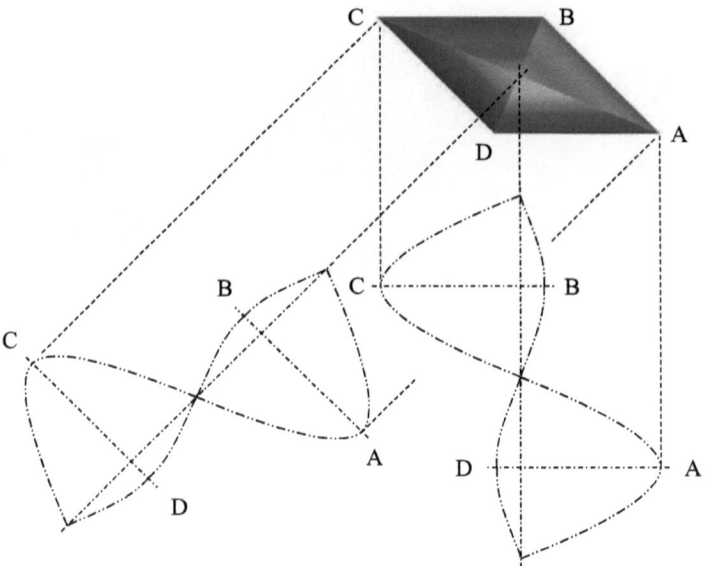

Figure 4.3 The elliptic-circular rhombus (ECR) and its expansion © Wladyslaw Chrusciel

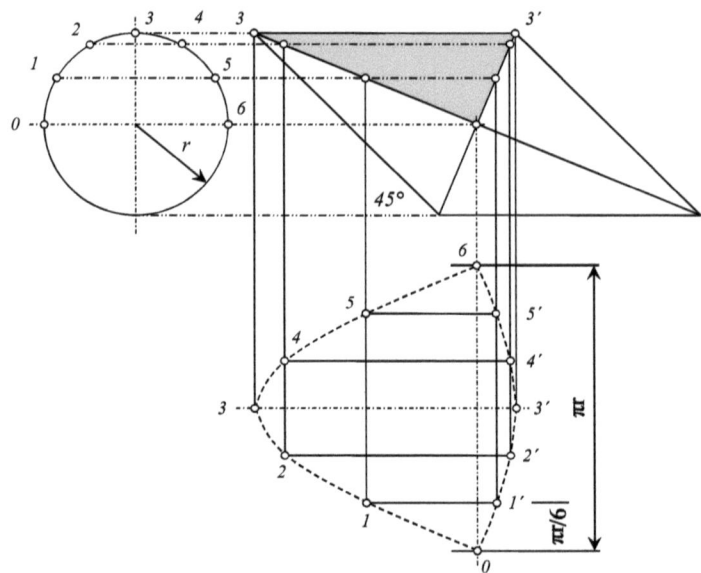

Figure 4.4 Practical development of the outer surface of the elliptical-circular rhombus (for the calculation is $\pi = 3.1415$); Points (zero circles) are only informative signs for the solution of the task and are irrelevant for the geometric construction

32

BIBLIOGRAPHY

This bibliography contains works cited in the text, which form the bibliographical panorama of the field of knowledge *geometric constructions*. Only a small part of these items is closely related to the division of an angle into N arbitrary equal parts, and some of the items relate to problem-solving strategies that relate to geometry in various ways.

The bibliography therefore contains studies on the relationship between science and geometry in particular.

[BIB52]   Biberbach L., Theorie der geometrischen Konstruktionen, Basel 1952.
[BRE51]   Breidenbach W., Die Dreiteilung des Winkels, 2nd. B.G. Teubner Verlagsgesellschaft, Leipzig 1951.
[CHR20]   Chrusciel W., Neue Bestimmung der Teilungsmethode eines beliebigen Winkels in N gleiche Teile nur mit Zirkel und Lineal, Hinterlegungsurkunde Aktenzeichen: 178/20/UrhebR/Roe, 12.11.2020, Rechtsanwälte Böttcher-Roek-Heiseler, Berlin 2020.
[DUS17]   Du Sautoy M., What we cannot know, 4th Estate, London 2017.
[EUK10]   Euklid, Elemente. Die Stoicheia, ins Deutsche übertragen von Dr. phil. Rudolf Haller, 2010.
[FRO19]   Froese N., Euklid und die Elemente – Die Entdeckung der axiomatischen Methode durch Euklid, Creative Commons Attribution – Share Alike 3.0, 09.04.2019.
[GSE87]   Giering O., Seybold H., Konstruktive Ingenieurgeometrie, Carl Hanser Verlag, München, Wien 1987.
[KLE24]   Klein F., Elementarmathematik vom höheren Standpunkt aus, Bd. I bis III, Berlin 1924-1928.
[MES60]   Meschkowski H., Ungelöste und unlösbare Probleme der Geometrie, Frieder. Vieweg & Sohn, Verlag, Braunschweig 1960.
[MIL86]   Miller M., Gelöste und ungelöste mathematische Probleme, Verlag Harri Deutsch Thun, Frankfurt am Main 1986.
[NEI99]   Neidnicht M., Quadrund – Patent Nr-DE39917218/39917218.1, 1999.
[PER17]   Permesser S., Du Winkeldreiteiler, Diplomarbeit, Universität Wien, Wien 2017.
[POL10]   Polya G., Schule des Denken vom Lösen mathematischer Probleme, 4. Auflage, A. Francke Verlag Tübingen und Basel 2010.
[RED18]   Rediske A. C., The Trisection of an Arbitrary Angle: A Classical Geometric Solution, Journal of Advances in Mathematics (2018), Volume: 14 Issue 02, pp.7640-7669.

YOUR KNOWLEDGE HAS VALUE

- We will publish your bachelor's and
  master's thesis, essays and papers

- Your own eBook and book -
  sold worldwide in all relevant shops

- Earn money with each sale

Upload your text at www.GRIN.com
and publish for free